Table of Contents

Chapter 0: Preface

Great, another book on a lean tool? Yes, in some regards, but something a little different. We think this one fits into a compact niche and offers a different approach that some may find useful for specific applications. If you are in need of developing optimal layouts to maximize space and flow (any type of flow) using a unique and methodical approach, this book can prove to be a fun, quick and informative read for you.

We decided to document the experience and learning accumulated over the years on this topic to share with others and also "get it out there", and to show others the capability. It walks you through and describes a systematic way to get the task of developing innovative layout options fast completed in a team, ideally event based format, in a fun and perhaps controlled chaos manner. It is written in a story like flow to make the reading a bit less technical, but also provides perspective and application summary sections to capture the details in the spirit of a field manual. While the book is grounded in a manufacturing setting, the concepts can be applied anywhere.

Why a shorter book one may ask? It is intended to be a quick read as we are all busy and on the go today. Doing a lot of reading, I myself prefer quicker reads to get concepts across but have some substance at the same time. We aimed for a practical overview with minimal fluff to get the outline and concept to you, the reader, while also providing access to resources for additional information as well as supporting tools virtually if desired or needed.

I personally along with a few team mates have applied it for many years in manufacturing settings and have trained many people on the concepts with a broad amount of successful outcomes. Personally, I have a passion for manufacturing and have grown up in it, and will always be an advocate for manufacturing in our country. I believe integrating technology and lean thinking with an emphasis on people will be key to our future competitiveness in America, and applied tools like this in my view are designed to help us improve, in this case with respect to space and material flow. I hope you feel the same. Be sure to check out my personal blog shown below and follow for access to additional content as well as supporting tools and materials related to L3P. Enjoy, keep learning, and always compete.

Brian D. Summerfield
www.briansummerfield.com
September, 2018
Published by TwoEighty3 IC LLC

Published in the United States of America

First Edition, 2018

ISBN: 9781520475370

Published by TwoEighty3 IC LLC

202 Main St, #190

Pine Bluffs, WY 82082

www.briansummerfield.com

Page left intentionally blank

Chapter 1: The Story

The book describes a mostly fictitious story line about a fictitious corporation, Dynet, and one of their sites in the Midwest US, Breckinridge, going through a pressing dilemma to free up some space to in source work, in order to walk the reader through the L3P or Layout 3P system. Names, locations, timing, and characters have been modified or created for the reader to follow a simple storyline and make the concepts a bit more realistic and easier to follow. The plant is facing a situation where they are struggling with growth and cost challenges, and are presented with an opportunity to in source production from other facilities and suppliers, but are constrained with space in more traditional batch flow environment.

The chapters begin with the story narrative, with each chapter then providing a perspective and application section to provide more insight into the concepts presented in the narrative. The chapter then ends with an action summary to provide a detailed list of steps or concepts laid out in that chapter.

The setting and characters, all fictitious by name but not by personality for some individuals, include the following for reference.

Dynet – 110 year old corporation that manufactures HVAC equipment, with three divisions including Mechanical, Thermal, and Controls.

Breckinridge – Midwest US site in the Dynet Mechanical division. Site was an acquisition eight years ago by Dynet. Lean implementation has been slow, culture is somewhat poor, and performance overall has been mixed. It is the second largest site in the Mechanical division by revenue and size.

Greg – Dynet Breckinridge Plant Manager. He has been at the site for over 10 years, first in operations management, then the plant leader role for the last 3 years. Greg is long time operations management veteran, having worked for a number of companies, and has a more traditional view of manufacturing. While he does support "lean", he does so with more reluctance and skepticism and feels compelled to at times.

Ben – Breckinridge Facilities Manager. He has been with the site for over 18 years, all in facilities and maintenance roles. Very opposed and outspoken to newer ways of thinking and doesn't view "lean" as useful, often referring to past attempts and failures at site.

Jeff – Breckinridge Manufacturing Engineering Manager. Started at another Dynet facility as an ME out of college and moved to Breckinridge two years ago as the ME Manager. Progressive thinker who is strong advocate for lean and new technologies.

Chris – Dynet Mechanical Divisional VP of Operations. He is Greg's direct leader and driver of the dilemma facing the site, wanting a solution for the business. Believes in lean and tools, advocate, but driven for a feasible solution first and foremost.

Harold – Breckinridge Materials Manager. Team member, who is somewhat reserved and sort of on the fence. He has been at site for over 10 years in materials roles, and is heavily influenced by Ben and often defers to Greg.

Maria – Site Quality Manager. Team member, who is a newer member to team in the last year, and a very progressive thinker. Can be quiet, as more introverted, but with right mix of people, provides good input.

Ace – Manufacturing Engineer. Less than a 18 months out of college, and very progressive in thinking. Part of Jeff's ME team and very supportive.

Ellen – Site EHS Manager. Also newer team member, been with site for about one year. Heavily influenced by Greg, her manager, but also provides high energy and good input.

Brian – Mechanical Divisional CI Support.

John – Mechanical Divisional CI Support.

Chapter 2: Set the Stage

"There is no way we have the space," said Greg, the plant manager, for what seemed like the 15th time today as we looked at the plant layout on the wall. "We just don't have the damn space to fit all this in here, and I'm going to have let leadership know," he added. The team was feeling the energy drain in the room. And to be honest, so were we.

The challenge was real. Find the space in an already jam packed factory to fit in not only a plant consolidation, with two value streams from another facility totaling 14K square feet (conservatively), as well as find the space for production of a new product expected to launch out of the gate strong and require two additional, dedicated cells at a minimum. Another 6K square feet, and with the addition, 20K square feet in total. The total plant existing space was roughly 95K square feet – the request was to free up nearly 20% of it. It was obviously a "good" problem to have for the business in terms of growth and opportunity, but nonetheless a challenge for the plant and the team.

Options were limited and so was time. Sound familiar? The team had to get creative and come up with some feasible options. Going back to leadership saying "it won't work" just would not cut it. The plant was under cost pressure already, and this was an opportunity to absorb additional work and level out the cost structure a bit and if anything, buy it additional time.

The plants change. The people and faces change. The seasons change. Even the situations and demands change. But all have the same theme in this scenario – develop design solutions using maximum creativity, sometimes under pressing time constraints. In this scenario above, in a plant in the American Midwest many years ago, we suggested a different approach to look at things truly differently. What we ended up using was a variant of a lean tool known by some as 3P, and refined today to be known as Layout 3P, or L3P for short. While not revolutionary, it's a tool we have developed a unique approach to and has proven useful time and time again in various situations around the globe.

I may not have suggested the approach there had this been a "lean" factory, with space and flow already optimized at peak levels. While I think there is always opportunity to continue making gains in space and flow in any facility, this particular one was fairly early journey in terms of lean, and a simple walk through revealed aisles too wide, "pockets' of free space strewn about, assembly cells with large footprints, machines very far apart, a sprawling materials warehouse, and in general, poor layout as could be seen through observation of motion and transportation waste. In other words, there was a gold mine so to speak of opportunities available to free up space and optimize flow.

I was an outside support resource, suggesting an approach, which along with John, another outside support resource with me, had used in various similar situations before with varying results, but was far from refined. "Why don't we try to run an event using a 3P approach and see what we can yield from it?" I asked. I was immediately criticized by Greg, the plant manager and Ben, the grizzled and grumpy facilities manager, who tagged on to Greg's frustration. "You have done a little work in this facility, but you don't have the experience we do. We have worked here for many years son, and know this facility and operation inside out," said Greg, the plant manager in a calm but frustrated tone. Ben piled on, "You don't even know what you don't know. We are absolutely out of space and obviously need to put in a capital request for a building expansion. I have quotes. It would cost about $1.3 million."

I added, "Exactly. I agree with you both. We don't know the facility as well as you, but we as external support could offer some outside perspective and new tools to help you, the experts, take a fresh look at things. To see how innovative we can be in our thinking and we don't have much time to respond back. Leadership wants an answer by the end of the week and its Monday afternoon. Even if you end up needing to justify an expansion, you will need some analysis and logic to back it up. This exercise could provide that support if it came down to it."

John, my divisional colleague, and I felt alone on stage at a talent show where we ran out of material. The room stared at us without a sound, and the plant manager, who had moved to the rear of the room now, was glaring at us ominously now as if we had insulted his grandparents. John tried to break the silence with a joke, but no one laughed. I interjected, "Look we are asking you guys to try something different. It might work, it might not. But considering the request, we don't have a lot of options. I don't think we want to go back to Chris, who was the VP of Operations, with words alone saying it can't be done. Unless there are any other ideas, we have a few days to pull something together. I would suggest we rally around this and give it a go."

The facilities manager got up and kicked his chair, then walked to the back of the room, grabbing and cracking open a soda along the way, standing next to the plant manager, and drawing sips. The plant manager said, "Dammit. As much as I want to kick the two of you out of here, we don't have a lot of concrete approaches here, and have been looking at this multiple times now. Let's give this crazy approach a try." The facilities manager nearly choked as he coughed mid sip. With the plant manager on board to give it a go, the team would align but not all would be exactly supportive.

Perspective and Application

Cooler heads would prevail that day and we would give the "radical L3P approach" a try. The rest of this book will walk through it as the L3P process is unfolded and explained. In perspective, as lean practitioners, we always face some type of adversity in people's minds and paradigms. In this case it was no different. The team here did not see immediate solutions and struggled with the concept of using a new approach foreign to them. Teams and us must keep an open mind and be willing to experiment and try new things, otherwise we will get the typical results. In this case, the report back to management would have been "we can't do it, it just won't work", with no strong supporting analysis or facts.

The situation described did occur many years ago, although the story lines and people have been changed to protect the innocent. This was the early days of building this Layout 3P process about to be described, and situations (quite a few) like this helped to hone what it is today. A high level process flow chart for L3P can be found in figure 1 in the appendix at the back of the book for reference. The fact is, if you have been, or ever are in a similar situation as this or even relatively similar, we believe this is the right tool for the task at hand. So keep reading and enjoy the flow.

Chapter 3: Introduction

About this Book

This book is a process guide to getting innovative layout work right in various environments and facilities. It will explain a streamlined and proven process to help you develop more effective and optimized layouts for facilities, plants, departments, or work cells fast with maximum team engagement. The outcomes will be a detailed layout and a plan for progression to a future vision that can improve flow, throughput, and space optimization significantly in organizations both large and small. This book will help you:

- Understand current state challenges and opportunities through a structured analysis
- Develop several innovative layout options fast and hone in on the best using a scientific approach
- Drive maximum team engagement through the process to increase buy in for implementation
- Create an iterative process to instill an evolving future state vision
- Create layouts to maximize material flow, throughput, and optimize space

About the Author

Brian Summerfield is a leader and follower, student and teacher, reader and writer. An engineer by trade, manufacturing leader by choice, and hardened lean practitioner by habit. With nearly 20 years of experience across a variety of industries and companies from Fortune 500 firms to leading private organizations as well as a number of SME businesses. Starting out on the shop floor in a foundry and progressing into machining and through the ranks, his training, experience, and practical application (failures and successes) has made him a subject matter expert in many tools while also honing his ability to lead the often most challenging cultural transformations through behavioral lean development. His ability to simplify strategies and tools and their practical, no nonsense application has proven effective from plant floors to offices to boardrooms. With a family history and roots in manufacturing, he is a passionate advocate for manufacturing and believes in the importance of it for America.

About the Publisher

TwoEighty3 IC LLC, a subsidiary of TwoEighty3 Holdings, is focused on intellectual capital in the leadership, talent, lean, learning, and technology spaces, with specific emphasis on these areas in manufacturing. It is focused on research, publishing, consulting, and producing value added content, services, and products to support knowledge management for manufacturing leaders and practitioners.

About the System

This book describes a system for doing effective layout work quickly and efficiently. The methodology we present here has been developed and refined over several years across a variety of industries and can work in manufacturing, health care, or even distribution environments. Within manufacturing environments, it can be applied in machining, assembly, and or any type of process operations. The method is particularly effective when taking a team approach, typically in an event format, but can also be used over a more extended period of time in more of a project oriented approach.

The system developed as an analytical technique for designing layouts and morphed into more of a lean event format in larger organizational environments, as there was a need to develop and deploy a more scalable version quickly in various areas around the world. The event format has been streamlined and can be done in a single day, although larger, more complex efforts are better suited for 2-2.5 days.

The book is straightforward and gets to the point. It is a quick read, and as such you won't find a great deal of fluff here as the format is the general flow of the application. I would suggest reading it front to back as it is a quick read, and then it can be used as a reference guide for future use. We seek to teach and inspire action.

Why this book? It is a way to capture and consolidate the material from over the years and present in a format targeting folks who have a need for a specific approach and tools, namely in this case doing layout work at the macro (plant or facility) or micro level (cell or other space), which can use a somewhat similar type of process. Some of the techniques are not new, like the 7 ways, but used in conjunction with other techniques to create a more formalized system.

We will say that it works. If you follow the process, it will get you where you need to be, which is developing sound layout options that can then be further detailed and put into an actionable implementation plan. We focus more on a team approach as having more inputs is where a broader innovation base comes from and a key part of the process. Input from many minds and eyes is always more powerful than just a few. Additionally, it builds more engagement to have several people, a coalition in other words, involved throughput the process, and this makes implementation go smoother and faster. If you combine all of these elements and follow the process, your chances of success will increase exponentially.

Chapter 4: Preparation

"We don't have much time to get ready," I told Jeff, the manufacturing engineering manager, and his team of three manufacturing engineers. It was Monday evening now, and we had to pull something together rapidly to get ready for a start the next day. The plan was to run the "event" from Wednesday through mid-day Friday with a report out to Chris and other leaders Friday afternoon.

John and I pulled some thoughts together quickly on a white board. It was late day Monday; we were drained from the mental jousting throughout the day with Greg and his team. We essentially wasted a day looking at an old layout multiple times and debating about possibilities. Nonetheless we had to pull something together.

We scribbled on the white board things we would need. I verbalized as I wrote, "we will need a couple of plotted layouts. Two current state "as is" and accurate as of today and two additional with nothing but the 'four walls'." "Four walls?" one of the engineers asked. "Yes, blank with nothing but the four walls of the facility on the paper. These will be the design canvases for the teams," I added. "And we will need all the machines, cells, benches, and other things in the plant cut out, so will need two layouts printed to use for cutouts. And scale of the layouts is important." I continued, "We will be doing something called 'paper dolls", where we use the blank canvas and then move the various machines and cells around on it to create different layout versions. It's a key aspect of this approach." The manufacturing engineering manager and his team were strong technically and understood lean concepts as well as being strong supporters. They picked up on what we needed right away and the ME manager said, "We go it. We will get it done by tomorrow mid-morning." With that, we confirmed we would start the event in the afternoon on Tuesday.

The bulk of the prep was in the layout plots and getting the cutouts for the paper dolls. John and I hung around for few hours in the evening to help get the cutouts done. The engineers cleaned up the scaling and made sure the current state layouts were accurate. We actually made great progress into the evening, getting all the layouts plotted on D size (A1) paper, and we got two sets of cutouts done, anticipating two teams. Feeling accomplished, we headed out for dinner and a few drinks.

Over beers later that Monday night, John and I sketched out (on napkins) some ideas on how to pull this off. "We will want to go with two teams, ideally 4 or 5 per team," I said. "Yes, and make sure we get a good mix of people on each team. We have some strong personalities that can drive group think here," John added. "We definitely want Jeff on the same team as Greg to help offset his constrained thinking. And make sure the facilities guy, Ben, is not on the same team as Greg, the plant manager." I agreed. And so it went on, with a listing additional supplies needed as well as a preliminary cut of the teams. All of which were compiled in an email to Greg and team back at the hotel before crashing for the night after a long day. I had no idea what was going to come of this, but at least felt we had the start of something.

Perspective and Application

Preparation is essential to a successful Layout 3P event, and should start at least a few weeks out ideally. Getting the materials accurately prepared is a key activity that should not be taken lightly, and likewise, putting thought into the teams and composition is a key aspect as well. Partnering with people closest to the facility and processes is always the best course of action for prep.

As will be pointed out later and also mentioned here, it is best to have multiple teams working independently with the same task in L3P. This drives more innovation, helps combat group think to a degree, and can also be used to drive a competitive spirit leading to more ideas. We have found competition to be an excellent driver of innovation in these efforts.

Chapter 4 Action Summary

Here are some of the key items typically needed for preparation:

- Current state layouts - at least (3) scaled layouts printed on large paper (at least D size/A1 preferred). If you do not have access to a plotter, you can utilize an outside service like

FedEx/Kinko's for a nominal charge. **Be sure to know the scale factor in the CAD layout**.

- Maintain at least (1) "As Is" for current state flow mapping
- Keep other (2) for cutouts of "paper dolls" (assuming two teams – need one "set" per team)
- The cutouts should be done ahead of the event and be cutout to the detail of cells and machines, so they can be moved around freely in different locations and orientations. Avoid cutting out large blocks of areas – this will defeat the intent of the exercise.

- "4 wall" layouts - blank slate layouts with only walls, fixed structures, or agreed upon monuments in place (this forms the design canvas)
  - Quantity depends on size of group and number of teams
  - Layout sheet per team for 7 ways exercise
  - Monuments should be discussed and agreed upon with team **during** the event
- Several sets of scissors, tape, different color markers/highlighters, rulers, and some length of strong or thin rope for measuring.

Who should be on the team? A wide cross section of people is good to have for out of the box thinking and free challenging and questioning. Core members should include:

- Operations leadership (**note**: *be careful about full participation of the top leadership. We have sometimes noticed that top* *leaders in these types of events can stifle team creativity due to any perceived intimidation or unwillingness for some team members to speak their minds freely.)*

- Supervisors or group leaders
- Team leaders
- Materials team members
- Lean staff
- Manufacturing engineering
- Maintenance/facilities
- Operators and team members (*a few at least should be selected - this should be viewed as very positive recognition. Use the membership as a reward or have team members "vote" on who they would like to have in the event as suggestions seen in past.*)

Supporting cast should include people from quality, planning, and procurement/supply chain. Functional staff can be invited as well although be aware their lack of knowledge of the operation could be a hindrance if not focused. If looking to build goodwill and get some true outside perspective, invite some people completely detached from the process. This can really be positive and actually appreciated if they challenge the status quo.

It is highly recommended that others, particularly some operators or other front line leaders who are not part of the core team be brought in later to review concepts and give feedback. They can be shown the layout options at later stages in a gallery walk, and can even be included in the voting process to be described later. It is also advised as a best practice to post the final layout concepts to come for a week or two and let the greater team know to take a look and offer some feedback. You can leave a flip chart and marker in the area or some Post-Its. Some good feedback has been achieved when doing this that was ultimately incorporated and it does a great job of getting the team more engaged and involved to drive acceptance which helps the culture.

Chapter 5: Beginning - Current State

It was a warm, mid-summer day, especially hot and humid in the Midwest. On Tuesday, we spent the morning with the ME team making final preparations on the layouts, cut outs, and gathering other supplies for the afternoon start. It was now Tuesday afternoon as the team funneled into the room, and John went over some brief ground rules and discussed the intent of what we are trying to do. I took note along with John that Greg was not in the room. "We want to come out of this with at least one viable layout option we all feel good about, preferably two," I interjected at an opportune time. I continued, "We are going to go through a somewhat methodical way of getting there, but try to trust us and follow the process as it may get rigorous."

With that John went over some very basic training we had thrown together the day before, combinations of past similar exercises and some content we pulled together. We answered a few questions from the group and then moved into the first part. With that, Ben, the facilities manager, chimed in with his thoughts. "We better get something together quickly. Leadership will want to see something fast and we can't overthink this and waste time. Greg and I are both reluctant to try this, so try to keep that in mind." You could feel his statements sort of deflate the room. Greg had still not shown up yet.

"Our first task is to better understand our current state and reality. We may think we know the factory very well, but it can be eye opening to study it in more depth," I said. With that we posted the current state layouts on the wall and asked the materials manager to write the data we had requested Monday night on a flip chart. "We have two major value streams right now in this plant. Both are Make to Stock (MTS) and one is roughly about 65% of total demand, the other is roughly 30%. The remainder is spare parts and miscellaneous products," the materials manager noted as he wrote demand numbers for the major product families on the flip chart.

"Great. Our job now is to break into two teams and each team take a value stream to trace through the facility, point to point in a spaghetti map fashion. We want to understand the true flow of the product through the facility, point to point. Each team will use a different color for their product family and work on the tracing as a team. If you aren't familiar enough with the actual flow, either go walk it to observe or ask for input from those who do. We don't need super precision here, but macro level accuracy to give us a better picture of the current reality," I pointed out.

On my last word out, I was interrupted by the plant manager, Greg, who had shown up just a few minutes previous, late to the event, "We obviously don't have the time to do this. We just did "value stream mapping" a few months back so why do we have to re-do it now when we need to be working on *future* layouts?" he asserted, riding a negative wave that Ben started earlier. "Well, good morning and great question Greg," I quickly answered. "This part of the exercise can seem redundant but we have found that it helps to level set the team and show the opportunities that do exist from a flow and space perspective. While similar, this is a bit different than value stream mapping, it is a bit more granular than that. It will also show us the reality of how bad our flow is and maybe how poorly we are utilizing space today," I replied.

Greg glared at me as he paced ominously in the back of the room sipping his coffee. I took no reply as a sign to keep moving and we did. Him showing up late, along with the low energy comments from Ben earlier and now him, already had the team off to a disjointed start, and was a sobering reminder on how leadership is so crucial to this or any type of improvement event. Leadership has a profound influence and can set the tone – either good or bad I reflected in my head. In this case, influential, highly respected leaders like Ben and Greg were driving us towards a negative start.

A great question from Ace, one of the manufacturing engineers, came up as to what parts to trace and broke my train of thought. "Follow the core component of the product for this study. In the case of a pump, follow the housing. If we had more time, we could also look at flow of the supporting components, but you will get a good idea of the flow following the main part," John replied. The teams proceeded to study and trace the product flows on the layouts.

I did a brief review of material covered in the training overview, on the '7 flows' to the team, stating, "When we analyze flow, we should be honed in on 7 aspects of it. They include the material flows, raw material, WIP, and finished goods, the flow of operators, flow of equipment, flow of information, and flow of engineering. When we think about the 8 wastes, we typically see transportation and motion waste, but could see examples of any." I pointed out that transportation waste deals with movement of material, whereas motion waste is typically associated with operator movement. With a quick review of wastes, we asked the team to identify wastes they recognized or specific flow related opportunities on Post-It notes next to the layout.

Once the point to point mapping was done, we moved to the next step. "We want to know the rough travel distances – use the ruler and measure off the distances on the map. Log them, and then use the scale to translate to feet. We will then do some math with the demand and batch sizes to get the rough annual distance traveled, which is often some sobering numbers," I concluded. At that point, Ben, the facilities manager spoke out loudly, "Enough of this! Greg, let's get focused and just go work on the capital request to expand the building. We did some investigation yesterday while these guys were wasting time cutting out paper and we can do it for about $1 million. This is a waste of time – let's pull the plan together and present it to Chris," he barked. For once, Greg, who was standing in front of the maps looking at the mess of spaghetti on them, took the other stance, and said, "No. This is truly a mess, worse than I thought to be honest. Let's see what the numbers are. If we do end up having to expand, all of this may give us the case we need to do so," he added bluntly.

The measurements and math ended up being sobering. The primary value stream travel distance totaled over 5K miles annually, while the other was actually worse, at roughly 6K miles. The maps showed a convoluted path of twists and turns, back tracking, and winding mazes, - and we only had traced the primary core component. The parts would add even more to the complexity. "This is crazy. We had no idea the flow was this bad and it's right in front of our eyes every day," said an operations supervisor. The room was quiet as they looked at the maps and the numbers on flip charts. But we had to keep moving. "We also need to look at space use. Value added versus non-value added," John inserted. "Value added space is space utilized for production or adding value to the products. Non-value added space is unused space, inventory, or spaces being used for non-value added activities," he added.

The numbers ended up being just as sobering – the plant in total was roughly 23% value added space. Not horrible, but shows the opportunity we have to improve it. It was definitely a good time for a break and some decompression. Being that it was late afternoon Tuesday, we decided to call it and wrap for the day and pick up tomorrow. This current state assessment is the first pivot point.

Perspective and Application

The current state evaluation is often a sobering exercise and is a great way to start L3P exercises. Some may view it as a waste of time or redundant at the beginning, but we have always noted that once team' s go through it, it does the trick in setting the table for the work to come and is a good wake up call. We call this the first pivot – recognizing how poor things really are and the opportunities we have in front of us. While it does take some time at the beginning, it is an essential part that should not be skipped nor glossed over. We have found it very helpful to drive more innovative thinking in the next part.

The current state point to point mapping is key and should be done for each major product family or value stream, in a different color on the same map. If there are multiple value streams, you may consider breaking the teams up and using two or more current state maps to trace the flows. You ideally will have a mess of a map – readable, but a mess for show, depicting your real material flow. Be sure to trace the flow as it actually is, and don' t "float" over racks or take the shortest path on paper. Trace the flows as it is today, even if going back and forth 10 times. If you are unsure of the flow, go to the floor and walk it, or consult folks who do know. Now is not the time to pretend things are better than they are – the current state gives us a true picture of reality. Once the flows are mapped, measuring the distances and converting will provide the material movement distances, which can then be annualized. The teams can then present on their maps and final numbers to each other. This requires the value streams and/or families to be clearly defined on the maps. Yes, this will get messy with several product groups but that is the point – the current state of your facility flow is more than likely a mess, worse than you think, and why you are doing this exercise. There is a reason this is sometimes called spaghetti mapping. An example of some current state maps can be found in the appendix in figure 2.

The value added space analysis is another key aspect, which like the flow mapping does for transportation analysis, gives us an indication of how efficient we are in utilizing our space available currently. Like the material movement, we should be trying to improve and optimize the VA/NVA ratio as much as possible as we go through the design process.

Completing all this completes the current state evaluation and leads us to the first pivot point, often a sobering view of reality. The pivot point turns to the future state and design process next and this exercise is to show the opportunity available and also set the sense of urgency in the team to develop better solutions. It should ideally inspire the team and show them we can do better.

Chapter 5 Action Summary

1. Create and plot a current 'as is' version of the layout

2. Identify value streams and core product families to

3. Review the 7 flows and 8 wastes with group – instruct group to record during mapping

4. Complete flow mapping by value stream on the current layout and summarize findings

5. Document distances and space as well as opportunities

6. Report out on the current state

Chapter 6: Design Objectives

As the team reconvened the next morning, Greg made a comment, "Yesterday was pretty eye opening. I had no idea our products were moving that much through the facility." I added, "It is a good exercise to go through with these because it does level set the team. The bright side is there are plenty of opportunities to improve flow and space." The team mostly agreed although the facilities manager, Ben, was still skeptical. We could see it in his reaction although he didn't say anything. It was an air of passive aggressive silence.

"We will move to the design phase now this morning. This is where things may get interesting and a little fun," I said, adding, "Our goal here eventually will be to go through multiple rounds of what is referred to as 7 ways. This is the point where we want maximum creativity and a focus on volume of ideas, no matter how outlandish they may seem." John inserted, "But first we need to talk about monuments, design criteria, and any go/no go items."

John asked the group if they knew what a monument was, while I prepared a blank flip chart. "Large, immovable objects," a supervisor shouted out. "That's real close. They are typically anything in the facility that is large, impossible or very hard to move, and if we did, would be costly and high risk," John said. "An example may be a forging press that is installed in a pit and also protrudes through a special roof opening. It would be very difficult and costly to move – likely a monument. Another example could be a heat treating system, which is typically heavily integrated with the facility and tough to move," I added. John summarized, "once we all agree on the monuments, we will lock them down on the layout, meaning they will remain fixed on our design layouts."

The team began talking about potential monuments and noting many, when John and I interrupted, letting them know that all things cannot be monuments. "We need to think outside the box with this, a monument has to be a legitimate monument, something we truly don't want to or can't move," I said. With that, the team identified and honed in on a couple of monuments which I wrote on the flip chart:

- Bathrooms and lunch room on east end of building. Would be costly to move and are actually in best location likely.
- Large Mori Seiki horizontal machine. Older machine, would be risky to move and costly.
- Mezzanine above assembly. Would be costly to move and require new inspection and permits. May be possible to take out.

"Great work, a couple of key monuments everyone agrees with. Let's highlight those on the design layouts," John stated. "We can now talk about design criteria. The criteria are important as it drives our design focus and is used in the decision matrix we will be building and using," I inserted. "What is most important to use in this effort? We want to clarify our primary objective," John said.

"We need space. We need to free up space," Greg commented. "We also have opportunities to improve flow," added Jeff, the ME manager. We agreed these were the top two, stressing space is needed and the primary driver. "What about cost?" Jeff asked. "Great question. With cost and time to implement, we include those as 'detractors', meaning they are design criteria but used to flesh out the most optimal cost and implementation time options," I answered.

With that the team agreed on the four criteria and the weighting (sum to 1) as follows:

- Space = 0.40
- Material flow = 0.25
- Implementation time = 0.20
- Cost to implement = 0.15

We then shifted focus to GO/NO GO discussion. "We need to talk about any hard requirements. These are binary in the voting matrix – either it is met or it isn't. Is there anything that is absolutely mandatory we want to evaluate in the designs that aren't covered in our criteria? I asked. After some discussion, the team agreed that no facility expansion may be one. While we agreed to note, we added that we may need the flexibility to adjust later if needed.

Perspective and Application

Moving into the future state requires reviewing and setting the design parameters. A first key step is talking through monuments, which are large, relatively immovable objects or areas that cannot be moved or can only be moved with great effort or cost. It is important the team understands what monuments are and where they are so they can incorporate this thinking in the design process. There should not be many but typically a few. Examples may include forge presses which have dug in pits and have vertical modifications with the roof, a heat treat operation with several large ovens in place, or a large mezzanine structure.

Avoid the pitfall of calling many things monuments. If a monument is called out by the team, talk through it and get agreement. With frank discussion, many so called monuments can be dismissed. Remember, we are looking for breakthrough improvements with the layout changes and as much as possible should be on the table at the start of this. The team needs to keep a truly open mind for ideas, regardless of how crazy they may seem through our normal lenses. If something does cost a lot or take significant time and effort to move, it should shake out in the decision analysis later.

Setting proper design objectives and then weighting them is another important step. Coming to agreement on the top objectives and then applying appropriate weighting will keep the team focused on design priorities during the rounds, and also provide the analytical inputs into the decision matrix. Top objectives typically revolve around space utilization and material flow, but could also include other flow aspects such as detailed material flow (raw, WIP, FG), people flow, and also other considerations such as information flow or even safety. While cost and implementation time are typically criteria, they are typically used as detractors, meaning the voting will seek options with best cost and best implementation time. The weighting after agreeing on the objectives is critical – it will be depended on later in the decision analysis. The sum must be to 1, with the higher priority items getting the higher weighting.

A final consideration is a review of hard requirements. This would be anything mandatory the design options have to meet – it's a binary GO or NO GO answer the matrix considers. An example here would be no capital expansion of the facility. The design options would be evaluated against this – GO if no expansion needed, NO GO if not. A NO GO could eliminate the option (if the requirement is hard and absolutely needed), or it can be used to lower the overall score of the option with a reduction factor, in essence applying a penalty.

Be careful, like monuments, when declaring hard requirements. There should be few if any, and they must be legitimate. Having too many monuments or too many hard requirements is a sign the team is over thinking them or being too conservative or even fearful of radical changes. Having too many will constrict creativity and limit design options before even starting. The goal of this exercise if innovation and bold thinking, which unlocks the real power of this process.

Chapter 6 Action Summary

1. Discuss monuments and agree as team on monuments in the current layout
2. Discuss design criteria, setting the objectives as team and then weighting them accordingly based on priority
3. Review any hard requirements to include in the analysis
4. Set up the decision matrix in preparation for round 1

Chapter 7: Design - Divergence

Having worked through the design criteria and monuments, the team was now ready to move into the design process. It was still early morning on Wednesday, and John and I felt we were relatively on track. Ben however was growing impatient, "It is Wednesday and we haven't even begun working on any options," he said in an irritatingly slow voice. "What are we doing here Greg? Are we trying to get fired?" he blurted. I interjected and said we are trying to find solutions. "It's time for round 1, where we get to put our creative hats on," I said, trying to divert the team's energy elsewhere, rather than Ben's oratory.

John gave a brief introduction to the 7 ways process as well as the paper dolls technique, which is a key aspect of the design rounds. I then divided the teams up, with 6 people on each team. "We want each team to go through the round and develop as many potential options as possible, using the blank layout and the paper cutouts," John instructed. I added, "Now is the time for creativity. We want volume in this round – as many ideas as possible. Don't worry about things that seem outlandish or crazy, develop a concept, and then move to the next one." John used the analogy of a funnel. "We are at the beginning of the process, the wide part of the funnel and we expect a large number of ideas. As we progress through the rounds, we will filter out the best ones to keep moving forward with," he added.

I allowed some time for the teams to get into their areas. Each team had a large table with the layout on it, and the cutout pieces positioned around it. We did a final run through to check the cutouts and ensure they were trimmed to enough detail to allow movement and design on the layout. When all looked good, I stated that once a team gets a good layout, let us know and we will snap a photo of it. We will then label it and print it out so we can post for review. They could then move to the next layout. "We will get 90 minutes in this round. Again, think outside the box or even better, there is no box. We want divergent thinking and quantity over quality in this round," I said.

With that, the timer started and they began. After 15 minutes, John and I checked in with each other, noting the teams were still struggling to get the first layout. A common hurdle at the start of round 1 is too much discussion, too much analysis trying to get "the perfect layout" the first try. "Less talking and more doing," John said to urge them on. "Get the first layout, someone drive their concept, then move to the next idea," I added. Unless the team is full of engineers and analytic types, usually after the first layout, the teams begin flowing with more ideas and concepts.

As was the case here, as at the end of 90 minutes, team 1 ("Greg's team") had 6 layouts, while team 2 ("Ben's team", yes we split them up) had 5 layouts. Not quite 7 each but the team's made a good push and we had 11 layouts in total which isn't a bad haul. After that barn burner of a round, I gave the teams a break while we worked with the ME's to print out the photographed layouts on

A3 paper and we numbered them. We then posted them on the wall, and asked the teams as they returned from break to organize their assumptions under each as well as trace the flow and any free space at a high level. "We want to do a gallery walk, so be able to explain each layout to the group so everyone understands each one," John said.

The group went one by one through each layout, while someone from the team explained them and answered any questions. "It's important we all understand the various options as we will be heading to vote next. So if any questions, please ask them," John added. After getting through the 11 layouts, we asked if any of the layouts were similar enough where we could merge any, something we called stacking them. The team did find two that were very similar, #3 from team 1 and #5 from team 2. The group chose option #3 from team 1, and then we collapsed the other layout under it, essentially giving us 10 unique layout options now.

"Some of these are crazy. We will never get these layouts implemented," Ben blurted, with his reliable pessimism. "Ben, please refrain from the negative comments. We said before this round was about creativity. We never said any of these would be final versions; we still have more rounds to go. What these can do is trigger some different thoughts and ideas," I added. There were definitely some unique concepts with key assumptions such as cell footprint reductions or retiring old equipment that on paper showed some good streamlining of material flow and provided enough free space in a few options for the planned additions, which actually surprised Greg and some others. Hell, it even surprised John and me a bit.

With the gallery walk complete, we walked the team through the voting process. "Each person will get a voting sheet, with the design criteria along the vertical side, and the layout options 1-10 along the top. In the matrix, for each criterion (rows), you get three votes. The best option gets a 9, the second best gets a 3, and the third gets a 1. The remaining options in that row get no vote. You then repeat for the second criteria, in this case, material flow. With four criteria, you should have a total of twelve votes on the sheet, three per row," I instructed pointing at a sheet as I talked. John and I answered a few clarifying questions, as this can be a confusing but important step. "This is a silent and independent vote as we try to limit bias. Turn you ticket into us when done," John said.

By this time it was lunch time Wednesday. It was actually a good break point, as a next step would be entering the ticket data into the Excel based decision matrix. While we were behind just a bit, we had wanted to get completely through rounds 1 and 2 today, and still felt we could. The teams were exhausted after a first day working through something completely new to them.

At lunch, John and I talked through the afternoon agenda. "I think Greg is coming around, but we still have some ways to go with Ben," I said. "He is one negative guy for sure but has a ton of knowledge," John added. "But I think the greater team is beginning to see the logic and potential," he said.

I had entered the voting data in the decision matrix. The results were automatically computed based on the logic. The top 3 layouts we were seeking, came out to be #3 with 43% (team 1), #4 with 35% (team 2), and #4 with 31% (team 1), in that order. I prepared the summary for the team to review when we started back up in the afternoon.

The group was lively after lunch. Even Ben was high energy and extra talkative. "Let's get after it," John said. "I pulled the top 3 layouts from the wall and posted them on an adjacent wall in order. "These were the top 3 from the decision analysis," I said. I brought the matrix up on the screen and showed them the breakdown. I continued, "Based on the voting and the decision criteria and weighting, this is how the ranking came out. We are interested in pulling the top 3 out of this round, which you see posted in the back of the room there. We want to do what is called plus/delta analysis of each one to learn what we like and dislike about each. We will use this learning to propel us into the next round."

Almost on cue, Ben interrupted us and began reading an email out load from Chris, the VP, to the site leadership team reiterating the importance and urgency of finding a good solution as quickly as possible for the consolidation proposal. Ben read the email louder with each sentence to emphasize, "If the Breckinridge team is unable to find a good solution quickly, we will be forced to go with plan B, which is the Tijuana site. While they have plenty of space, I am uneasy of their quality performance and ability to integrate these value streams. I would be extremely disappointed if we can't find a way in Breckinridge. With the cost pressures and stunted growth domestically in Breckinridge, it would not be an ideal scenario for the site." Ben threw his empty coffee mug against the wall, "Dammit Greg! We have to stop playing around here. Let's stop these childish games and just put together a capital request for a 5,000 square foot expansion on the back of the building and we can fit it in."

Greg paused and then asked, "Ben, we don't even know if adding the expansion will allow us to fit the new product lines in. And if we do add the space, we would just be cramming the new value streams in here, with no improvement to flow or space utilization. I want to stay on this path and see where it goes. We need an analytical solution and I think this will get us there." It was breakthrough moment for a plant manager at a critical inflection point. He could of said let's stop this and just go for a quick and easy, but risky and costly solution. But he stood his ground and showed the team he was committed. I looked at John and we knew Ben, who was stunned, would be less of an issue going forward. Ben stepped out of the room to collect himself. Greg followed and they had a chat in the hallway.

After the group collected themselves, we continued with the

plus/delta review. The team worked through each of the top 3 layouts, noting on flip charts beneath them what they liked and disliked about each, having some great discussion along with it. This would provide fuel for the next round, as ideas were already being tossed around in the banter. This is what we call the second pivot point.

Perspective and Application

Round 1 is about innovation. We use the "7 ways" philosophy to push the teams to drive for 7 unique layout options. There is a lot of mystique around the 7 ways and Toyota and Shingijutsu lore. We are not going to give you a history lesson on the origins of the seven ways because chances are, you likely don't really care. If you do, you can do some Google searches and you can read all about the various accounts of who created it and debates about who is the best at it. The 7 ways is simply an innovation driver in our experience and a key part of the L3P process. While each team may not come up with 7 options, having them push for it drives the urgency and creativity. We want **divergent thinking** and **volume** in this round, not quality. Crazy ideas are welcome. "Try storming" versus a lot of talking should be emphasized.

We have elements of the forming, storming, norming, and performing model going on in these L3P events. Round 1, the teams are forming. People are professional, reserved a bit still.

This is part of the reason the first layouts take a long time, in addition to over analyzing. Through the first round and the later rounds, we look to do at least three, will take the team through the rest of the model, and you can typically see the dynamics change in the teams. By the later rounds, they are typically humming. Facilitators should be on the lookout for group think and certain people dominating the layout thinking. We want group interaction and discussion.

The timing of the round is a key aspect. It harkens to the 'Escape Room' phenomenon. If you have done one, you know that you are locked in a room with your group and have a set period of time, say 60 minutes, to escape by solving problems and puzzles. The time aspect adds a great sense of urgency and really drives creativity and ingenuity. The timed rounds of L3P are designed in this way, and it often leads to some creative ideas. It also forces communication and collaboration, which are key to developing breakthrough concepts.

Teams of 4-7 are ideal for this exercise in our experience. More than 7 and you get disjointed sub-groups and a few people leading the pack. Less than 4, it's just not enough eyes, ears, mouths and minds. If possible, have multiple teams; say two or even three if possible. You get different thinking in different teams, and driving some competition between them is good. We have used things like rewarding the team with the most layouts in round 1 with candy or something to having the "losing team" (least layouts) have to sing a song for the group. Teams like to win and not lose. Use that friendly competitiveness as leverage.

The gallery walk is a key step in ensuring the group understands all the layout options. Having clean photos printed out, preferably on A3 size paper or larger to hang up is key. The photo method sacrifices some scale difference quite often with the table layout size, but it is the quickest method we have found and does the trick. The gallery walk segues into the silent voting process which is facilitated with the voting tickets and some explanation. This data is then entered into the decision matrix which computes round results. Examples of the voting tickets and the decision matrix tool are shown in the appendix in figures 3 and 4 respectively.

We want the top three options by score so we can do the plus/delta analysis on them. Here the group talks through in detail the likes and dislikes on each of the top three options. In doing so, some healthy discussion and new ideas emerge. This is the second pivot point and a critical one. This information feeds right in to the next round, round 2, also known as the breakthrough round as many unique concepts from round 1 are often merged to form some very creative ideas. The teams are typically fired up heading into round 2 with ideas and new thinking from the round 1 discussion.

Chapter 7 Action Summary

1. Explain the intent and overview of the 7 ways process

2. Review the concept of "round 1" and the desire for volume over quality

3. Execute a timed round 1 in teams – if time; go for 90 minutes, no less than 75.

4. Plot the layouts, number them, and do a group gallery walk

5. Have each person do the independent voting on tickets

6. Use the ticket data to enter into the decision matrix, identifying the top 3

7. Do the plus/delta review of each of the top 3 with discussion

Chapter 8: Design - Convergence

The team was primed and ready for round 2 as John and I prepared the room for the next step. We moved the results from round 1 to a back wall and left the top three plus/delta results in clear view of both teams. We asked both teams to confer one last time on those outcomes and prepare for the second round. "This round we will go for 60 minutes. In that time, we want each team to again strive for as many layout concepts as you can, but pay a little closer attention to details and use the learning from round 1 to hone ideas. It's balancing volume of ideas still with some quality and details," I stated.

Greg asked if we still were looking to do seven layouts, and I answered, "If you can get seven quality layouts, that would be fantastic. Shoot for more than two to three at a minimum that are unique in nature. If you get more, than great." John set the timer as the teams prepared their tables. At the last moment, I switched up two team members from each team, moving Harold (materials manager) from team 1 to team 2 in exchange for Maria (quality manager). I noticed Harold was a bit outspoken on team 1 in previous round while Maria was a little quiet and was looking to shake things. When the teams were set, we began.

Round 2 brought a greater focus from the teams in the afternoon. You could sense they were more in the norming stages, as they jointly focused on their first layouts, which melded different concepts from round 1 and discussion points from the plus/delta. Both teams were quick in completing their first layouts in roughly 15 minutes and proceed to making slight, but useful changes to create alternate versions. A second round always brings some innovative ideas. After the 60 minutes, team 1 had four layouts, while team 2 had three. As with the first round, we worked to get them printed, numbered, and posted on the wall. We then had the teams do a gallery walk to review each one.

Both teams had some good ideas but also had some similar concepts in terms of flow. "I like how we both have layouts that emphasize a linear flow in the facility, with raw material coming in the west end of the building where we having shipping today, and then moving to the east side of the building where we aren't even using some of the docks and doors there. We actually are using the north end of the building for receiving today," Harold commented. "It's a complete flip of the factory, but I agree the flow is much more linear and it frees up a lot of space," Greg added.

Completing the gallery walk, the teams repeated the voting process after we handed out tickets. John said, "After you hand us your tickets, feel free to take a break while we compute the results." It was early afternoon now, and we felt like we were in pretty good shape. Our goal was to make it through a third round of voting by the end of the day. But more importantly, ensure we are honing in on some good concepts, which we felt we were.

The decision matrix was populated for round 2 and the outcomes were fairly clear. There were two clear winners, a distant third, and then the other four. Option #1 from team 1 (57%), option #3 from team 2 (43%), were the clear winners, while option #4 from team 1 (31%), was in third. No other option scored above 20%. The top two options both utilized the linear flow the teams discussed earlier, and the top three all showed some good space utilization. While not quite enough for the new value streams, it provided enough space to work with and solutions to the space challenge.

We pulled the top three options off wall and moved them to an adjacent wall. We then asked the group to walk through the plus/delta again, emphasizing to take some time and really discuss out the pros and cons, as this would be key for the next pivot point. You could tell the teams were getting burned out as they slumped in chairs and in general had a bit harder time focusing now. Teams typically get tired mentally after the first few rounds; this is a whirlwind type of effort. We pressed them on, "We know you guys are getting tired, but this is where the best ideas start to form. You guys are on to some key aspects that could really work as we refine them," I said.

The discussion during the plus/delta for round 2 was lively and the teams generated some great pros and cons for each. More ideas were flowing as the discussion commenced, and I encouraged them to keep these top of mind for the next round. This is the third pivot point – the outcomes of round 2, which will feed into the next round.

After completing the plus/delta, it was around 6 PM, and we decided to call it for the day as the team was cooked and quite frankly, so were we. "Get some rest and let's be ready to roll at 7:30 again tomorrow with some fresh thinking for round 3." I said. John added, "Great work today and hopefully you guys are beginning to see some clarity in the possibilities."

Ben approached us and we prepared for some more verbal jousting. He said, "You know, I think we are on the right path here. I apologize for my outbursts earlier. I get fired up at times and want to do the right thing. I have been here a long time, and many of these people are like family. I want what is best for the plant in my last two years before retiring. I'm seeing the logic in this now, and even though it is tiring as hell, am feeling better as we proceed." John and I looked at each other before I said, "As long as you aren't throwing anything, we should be good. Glad you are seeing the approach."

John and I debriefed that evening over dinner, and strategized the next day's agenda. We had planned for a group dinner Thursday evening, hopefully to celebrate coming out of this with a viable solution we can present on Friday. We discussed how the teams focus was sharpening even though the mental exhaustion form the multiple rounds were there.

We arrived early and prepared the room for the day. The team began funneling in around 7:30 and we got things going with a quick energizing team building ice breaker to shake things up.

Ben was noticeably absent, and we checked with Greg who was not sure why he was late so decided to call him. We carried out the team building exercise, a quick version of build the tallest structure with three teams, and it brought some good laughs and team competition. Ben unfortunately showed up after it was over.

Getting back to the event, John said, "This third round will go for 45 minutes. This round emphasizes refinement. If you recall the funnel analogy from earlier, we are in the lower part now and should be filtering out the best ideas. We have had two rounds now, a total of 17 layout concepts, voted out the best ones, and some great plus/delta discussion. You should be well equipped now to develop a few more of the best options." I added, "We want you guys thinking near term and longer term in this round. Take the concepts and look at what options are within the next 12 months, and also look at developing concepts for longer term, say 18-36 months."

With that the round began as John set the timer. Round 3 shows exhaustion on the thinking but also shows determination, as the teams have honed in on what they are seeing in principle and try to refine to the point of perfection. By this time, the teams have likely isolated a few key themes for the layout, and are typically trying to address some of the "deltas" or incorporate some of the "pluses" in the layouts. But you also see some new and innovative ideas emerge in round 3, and this usually leads to some new, radical thinking. It was good in this case we had begun round 3 in the morning with a refreshed team.

At the end of 45 minutes, we were getting later in the morning. We worked to get the layouts printed and posted. Team "Greg" developed three good options, while team "Ben" had two. You could see how the layout thinking had evolved through the three rounds, and now the flow was looking streamlined in all of them, as well as free space utilization looking more effective. The team was also able to do so without any extensive facility modifications in most options. We knew we were getting close. We gave the teams a break while we compiled the results and prepared for the next step.

We were just getting kicked off after break, when Ben emerged into the room late again. "Chris wants to have a conference call and review progress. He has heard we don't have any solutions yet, and since it is Thursday wants to know what the hell is going on," he said. John and I looked at each other. "Who did he hear that from?" I asked. "It doesn't matter now. He wants to have a conference call with the plant staff and you two at 1:00, and I will have the conference line open. He will expect a full report out on what we have been doing this week and what solutions we do or don't have," he boasted. John and I then knew Ben was likely the one who had reported to Chris and drove this. Apparently, Ben had not come around. Maybe it had been a rouse. In any event, we were stuck. It was 11:20 and trying to move into a gallery walk and voting process before lunch and the call just after would be tough. We also had to now prepare a "report out" on progress.

John and I had the teams do the gallery walk, and while they did, we prepared a couple of flip charts to summarize what we have done the last few days. We then repositioned the previous rounds work on the walls in order of progression so we can walk the callers through it. "What are you doing?" Ben asked. "They are expecting a formal PowerPoint presentation, not some things on the wall to look at," he added. "Ben, we think it would be more impactful if we used the actual work flow and walked them through it. We can use the call in, but we can also utilize the video conference feed to show the room," John stated. "Suit yourself. I know what Chris is looking for and it's not that. When he sees you guys drawing on flip charts and post its with markers for the last few days, and having no good solutions, he will stop this and let Greg and I get down to business and save the day," he said brashly. At that point, we knew exactly where Ben stood.

The teams finished the gallery walk as we finished up the preparation, and we broke for lunch around 12:15. I wasn't really concerned about the presentation, but more concerned about the interruption in the flow of the event. The teams were onto some good thinking and this could disrupt the creative flow. We went through who would speak and what to say over a working lunch to be ready for the 1PM call.

As Ben was quick to get the call in line fired up at 12:55,

we got the video conference working to show the room. We exchanged greetings and shared how the teams were at an important point in the design process; gently emphasizing this could be disruptive. Chris and the other leaders on the call understood, and asked us to quickly walk them through. John started off explaining the L3P process logic at a high level, and then we had Greg (yes Greg) walk us through the current state and findings, and then through round 1. We then had Harold walk us through round 2 and into round 3 where we stand today. "A solid understanding of the current state and opportunities, defined design criteria, and now twenty-two innovative layout options considered. Round 3 netted five unique concepts and we are about to vote on those to see what the best ideas are to proceed with," I ended the summary with. Ben sat in the back smugly and didn't say much at all, but you could see his frustration as the summary report actually flowed quickly and well. Chris asked a few questions on the space and potential to fit the additional business in. We clarified that it looks feasible, and it's just finding the optimal layout to do that, maximize flow, and also provide some future growth flexibility as needed, while balancing cost and implementation time.

Chris said, "This looks good. Nice progression. We were getting concerned based on what we were hearing coming out of there but I think this all makes sense. Get on with it and keep going. You still have to have a final solution by noon tomorrow. I trust everyone is aligned and on the same page now." We wrapped up the call and then tried to get back in gear to continue. It was 1:45. Disaster averted, but still a roadblock that burned a few hours today navigating.

After giving a quick break, we moved to the voting process for round 3. We made sure the teams were calibrated from the earlier gallery walk, and tried to put the report out behind us. After some brief questions and answers from a few, the team proceeded to vote on the five options. As before, the tickets were handed in and John entered them into the round 3 matrix.

Round 3 voting yielded two clear results, option #2 from team "Greg" (52%), and option #1 from team "Ben" (46%). Both of these options were fairly similar, with some subtle differences. The other three options were at 23% and lower. We pulled the top two options to the front wall and guided the team through another plus/delta discussion. We emphasized some deeper discussion here as this would be the fourth pivot point; we were close and wanted the next and final round to be the home stretch. This carried us into midafternoon. The afternoon needed to be the final run and then building a game plan to get to the future layout.

Perspective and Application

L3P after round 1 becomes a grind. It's the subsequent timed rounds that drive a filtered view of the layout options and helps the teams hone in on the best solutions in driving from divergence to convergence in progressive rounds. Round 2 is a bit less time than round 1, and round 3 is a bit less than round 2. The flow of each round is very similar: design round, gallery walk, voting, decision analysis, and then plus/delta. It's the pivoting and building off the information, ideas, and discussion of the previous round that really refines concepts and layouts. An example of a later round paper doll layout is shown in the appendix in figure 5.

Altering some other variables can have some positive effects. Mixing up some team members can help shift dynamics. Altering the times of the rounds is another, maybe required if under time constraints. We have also experimented with playing certain types of music in the background to alter the team's mood and drive more creative thinking. You may also be limited on the number of rounds, although we suggest doing at least two rounds, preferably three. The beauty of this is the rigor, which can exhaust the teams but if followed, always yields the most optimal solution the teams see together and agree on by consensus. Mixing in some energizing ice breakers to exercises can help infuse some energy and also break up the rigor. Being creative is a key part of good facilitation.

Chapter 8 Action Summary

1. Prepare for and execute round 2 designs – go for 60 minutes if possible, maybe 75. Focus on balancing volume of ideas with quality and refinement.

2. Round 2 layout posting, gallery walk, and then independent voting with tickets

3. Decision analysis and results for round 2 – top 3 options

4. Plus/delta debrief on top 3 and learnings

5. Use that information to pivot into round 3 design – go for 45 minutes if possible

6. Repeat the posting gallery walk, voting and decision analysis to yield the top 2-3 options

7. Can select the top 1-2 here if limited on time and move to refine them as a group, or move into a final round 4.

Chapter 9: Final Design

We were running a bit behind due the midday presentation we had to throw together and pitch. The group had two good options from round 3 that while somewhat similar, had some subtle differences. The top one, option #2 from team Greg, won out in terms of material flow, cost, and time to implement. The second one, option #1 from team Ben, won out in terms of space utilization, meaning the voting indicated it utilized the available space more efficiently and actually provided some free space for future growth. The teams had worked through the pluses and deltas of each. We decided to do one last design round to see if we could come up with best alternatives.

At this point, we merged the teams together to work. "Our goal here is to develop the best layout for the short term, say the next 12 months, incorporating all that we have learned up to now. We will then pivot off that and the learning to create a longer term option in the 12-36 month time frame," I instructed. "We will get 60 minutes for this and now is the time for details. We are not looking for multiple options – we have in theory exhausted the majority. We want to incorporate the learning and ideas into these final cuts," John stated.

The groups pulled the plus/delta info closer to the table and they prepared the cut outs. We reconfigured the room a bit to allow the larger group access to the table. We then gave 15 minutes for the team to discuss strategy before we started the clock. "Don't deviate too far away from the top two from the last round. Those have emerged at the top through the process for a reason. Leverage them and build on them," I added.

With that, the group set out as the clock started and began orienting the layout in a similar manner as the top option from the last round. They discussed some details, and spent some time working on aisle structure and traffic flow, which become more crucial in the later rounds. They discussed loading docks door access, and ensuring safety for people moving within the plant. The EHS Manager, Ellen, provided insight on door access, panels, and size as well as location of man aisles. We provided some input as far as limiting for truck traffic on the shop floor and keeping it limited to the docks and warehouses, primary reason being safety.

At the end of the clock, the teams were still working diligently. They had created the near term version and were still working the longer term vision. We gave them an extension of 15 minutes to finish up, with John stating, "Our goal this week is the near term option. Let's ensure that is detailed out and where we want it. The longer term vision can be conceptual and something you guys refine after this event." The team turned back to the short term option and continued reviewing to finalize it.

The short term option locked in, the team worked to tape down the paper dolls. No photo was needed as it will be the last option in this event. They did the same for the more high level longer term option, which the team got hung up on as far as a capital expansion. With the short term option locked down, we posted it up on the wall for review.

"As we did with the current state at the beginning, we want to trace the material flow through the facility by value stream using different colors, and also using a highlighter, highlight free space in yellow," I stated. "This will give us a good visual representation of the flow and space utilization. Be as neat as you can as this is the final, actual layout. We will need to be able to read this next week when it is being put into CAD," John mentioned.

The team traced the flows through neatly and also highlighted the free space yellow, and the newly added product families in green. This provided a very visual layout that when compared with the original current state side by side, showed some true change and improvement, at least on paper. The team reviewed both side by side making some comments, and talking about the transformation. "I have to say, I was skeptical at the start. But this process, as tiring as it is, really guided us to a pretty good solution. Yes we have some tradeoffs due to time and cost, but I think we all are pretty aligned on this final option," Greg stated.

Ben went on to add, "I agree. I've worked here for 18 years now and have been accustomed to our current layout. I had trouble seeing how we can change it to make this work but this opened my eyes a bit going through this. Thanks for putting up with my negativity early on." John and I both added almost in sync, "This is a new process and we all learn together each time we do it. We are still learning how to refine it, but really like the outcomes we have seen. You guys were amazing in grinding through this but we still have some work to do. We now need to layout the key milestones for the moves and then sequence them, almost like building a project plan."

It was getting late in the day, around 5:30 PM. While we were behind a bit, John and I agreed to call it and bring the team in early tomorrow morning at 7:00 to begin the planning and prepare for the report out at noon, which gave us about five hours. We had planned a team dinner for this evening and wanted to ensure the team had the opportunity to decompress and enjoy a nice meal together.

Perspective and Application

The final stage of L3P is using the learning from the current state and previous design rounds to refine a final solution. At this point, we typically merge the teams (if there were multiple) and have them work together on a unified final version or two. At this point, the teams are typically exhausted, grinding through the multiple rounds designing and analyzing anywhere from 20-30+ layouts. You can rest assured that while they are exhausted, you have also explored an exhausted nearly all possible alternatives for the layout.

The final stage is also where more diligence and details have to be driven. If you think of the funnel analogy used earlier, we are now at the narrowest point. It isn't about multiple options in the final round, but using the learning from the pivot of the previous round to develop an optimal final solution. Aisles, material storage and flow, traffic, safety, door and dock access are all typical things that while discussed maybe in the last round, now have to be talked through in some more depth. Details matter more at this stage in the design process.

Once the teams have locked in the final design, tape down the paper dolls using clear tape. This will become the final "working version" and it must be legible for someone to put into CAD later. Hang the final version on the wall. We like to put it next to the original current state to give some perspective. Then have the teams trace out the value stream flow and identify free space as well as any new additions with highlighters. Then have a quick debrief with the team to get final thoughts on the layout next to the original current state. If there is time and desire, you can also measure out the transportation flow distances as done with current state earlier, and also measure out the Value Added and Non-Value Added (VA/NVA) space. This can sometimes be a good "analytical" comparison, in addition to just the visual.

If the team developed a long term option, this can be taped down as well, and the same process repeated, with material flow traced out and spaces highlighted. In this scenario, we had the team create short term and long term options. Other times, we advocate creating an 'Option A' and 'Option B', in case A does not work or is not approved, and then you have B option as a backup. We would normally decide on the path for this during the event, although typically the time to implement is a driving factor.

Chapter 9 Action Summary

1. Merge the teams into one group (if there were multiple teams).
2. Prepare layout table and space for one team.
3. Pull top options and plus/delta info from round 3 near the team.
4. Give the team 10-15 minutes to strategize.
5. Set timer for a 60 minute round – stressing to team we want final design option with the details.
6. Lock in and tape down the paper dolls on the final version.
7. Hang on wall (can be next to current state) for review and then trace out material flows by value stream and highlight free and added spaces. (Bonus: can measure out distances and VA/NVA spaces for comparison of improvement analytically).

Chapter 10: L3P Project Planning

It was Friday morning, last day and a report out to leadership at noon. The full team had joined for dinner the previous evening, and it was really helpful to decompress together and do some team building. It was just about 7:00 AM now, and John and I were making some last minute preparations ahead of the kick off. "What's the game plan this morning? We don't have much time," Ben inquired as he overheard us chatting. I nodded to acknowledge and then we asked everyone to gather around so we could start. "Today, we have a report out in about five hours. We have to set at minimum a high level project plan with some ballpark budgetary numbers and a rough schedule. We also have to prepare a report out so we can explain the outcome and summarize in a way that is coherent and get alignment on direction," I said.

John and I surveyed the room. It was a wreck; it had looked as if a gang of bandits had ransacked it the night before. Layouts, flip charts, Post-It's were scattered about the walls. Cutouts were strewn about the tables and on the floor. The back table was checkered with coffee cups and some trash from snacks and crumpled up paper. As we would come to see over the years, this was a mark of a good L3P session. It also meant we were near the end and could not be focused on cleaning up now.

"We need to essentially build a high level Work Breakdown Structure, or WBS next," John stated. I added, "We will take about an hour to review the near term layout and identify on Post-Its all the critical steps or milestones to get from where we are today to the future state layout." We instructed the team to stay at a "mid-range" level on tasks, meaning not too high but also not to the task detail of calling a contractor. "Don't think about timing or sequence right now – just identify the tasks or milestones and get them on the wall," John pointed out.

Ben, the facilities manager, and Jeff, the manufacturing engineering manager, began leading the exercise, which was encouraging, especially seeing Ben, who had been a challenger early on, take the lead. They began reviewing the layout and blurting out some tasks while others scribed. "Electrical upgrades in the new warehouse area. Move the current milling machines to the east side of building in current open space. Ensure foundation is sound or if pads are needed," Jeff and Ben took turns rattling off, as Harold and Greg wrote. Others chimed in with thoughts too and the team did a good job in roughly 45 minutes identifying all the key milestones and getting them on the wall. More importantly though, we could feel the team was in performing mode now.

"Next step is sequencing. Think about what has to happen first, second, third, and so on. Consider what key dependencies are, which are tasks closely linked to other tasks, decisions, or events. Consider what can be done in parallel together. We are slowly building out a project timeline," I instructed. With that the team began reviewing the tasks on the wall and Ben began numbering them. Jeff and Greg then began moving some of the key early tasks to the left, and then they began to sequence out the events. After about 30 minutes, the group had a fairly cohesive, sequenced layout which they placed on some butcher paper that was put on the wall. They connected the hard sequenced dependencies with arrows, while parallel tasks were stacked on top of each other.

"Looking really good guys. Let's take a quick break since we are making some good time," John said. During the break, John and I chatted with Greg as he asked about the next step. "Developing rough timing for each task and an owner is next. We then need to try and develop a rough cost for the tasks, as best as we can, with understanding that we will need to get more detailed quotes later," I answered.

After the break, John and I quickly summarized what we had explained to Greg, and the teams honed in and got after it to estimate timing and an owner for each task. It was approaching 9:30 AM and we looked to be in pretty good shape, having a rough cut timeline developed. We could then turn attention to budgetary numbers.

"At this point, let's split the team up. A couple of people, including Ben and Jeff, will keep reviewing the work tasks and try to put some estimates on them. The rest of the team will begin preparing the summary report out," John said. With that, we divided up the teams. Ben asked, "Can I bring a few of my guys in to help with these estimates? My electrical guy will have some good insight on the power costs, and my lead mechanical guy will provide some good estimates on the moves." John and I looked at each other, and quickly nodded in unison, "absolutely." John and I also discreetly high fived each other as Ben turned around, noting how far Ben had come in the last 48 hours. He was literally beginning to become a leader in this event. This would prove to be critical in the post event actions.

The report out team split into fragments, with some beginning to organize the room a bit, some others posting the layouts and flip charts in order on the walls so we could walk through them during the report out. We began to take some good photos of what was on the walls so we could prepare a deck for those who could not video in or could not attend the report out. While they did this, Ben and his team continued to make head way on the cost estimates. The team was making great progress and things looked to be coming together.

Around 11:00 AM, an early lunch arrived and the team continued a semi-working lunch to ensure the progress was continued. The rough numbers were in. "It looks like in all, we are looking at about 6-8 months to execute, and the total cost is projected to be about $360,000 to $400,000 which is far less than the $1MM building expansion we had been preparing," Greg stated with a sense of pride. "And it looks like not only will the new product lines fit, but we will also have a little breathing room space wise for anything in the future. And just looking at the material flow, it's a substantial upgrade," he added.

We finished up lunch and noon was approaching. The room was fairly clean now, the report out ready, and the team was chatting. John and I packed up our gear and prepared to head to the airport. By all intents and purposes, it was a rough week that had a crazy start but ended fairly well. We felt pretty good about the outcome. I shook John's hand, and said "nice work, glad to have you here. Not sure if one person could have pulled this off alone." John nodded, as he dialed into the conference line and we fired up the video rig.

Perspective and Application

If you are wondering how the final report out went, it went well with some good questions and praise from leadership. The near term plan was approved, and the team's task was to refine the plan next week. Jeff was assigned by Greg as the project lead, and Ben and team would support. Leadership and team felt good about the outcome and optimistic about the plan execution.

They definitely aren't always this smooth and some take a wrong turn and don't end up as clean as this one described. It's typically not the process alone that falters if things do go sideways, but a combination of factors and sometimes people, such as "Ben" described in this book. Ben and Greg both came around in the end. In some cases, folks like them don't.

The project planning session is the tail end of the L3P effort and a key part of the exercise. It is where the final layout option is planned on how it will be executed and brought to reality. The L3P event turns from the creativity driven exercise into a more detail oriented project planning session, with the buildup of a project timeline and rough cut budget. The team identifies all critical tasks and milestones first, and then works on dependencies and sequencing. The final timeline step is estimating task times and placing owners on them. The timing will be assumption driven – shoot for a range in timing and confirm later with detail. The team should also pull the risks and assumptions noted for the final layout option, and place those with the project timeline.

The last step is estimating costs. This will be very high level, but provide at least a conservative (shoot for a range – high and low) look at the cost of the transition. Here, it may be necessary to bring in some outside expertise that maybe wasn't on the team, perhaps maintenance team members or other personnel familiar with the facility. Get the right people in the discussion so some level of accuracy on costs (and even the timing) can be developed.

Finally, tie it all together in a package for summary and presentation. You need to "tell the story", showing the layout options and how the team progressed there. Those outside the event may not understand the rigor of the process or that you worked through 20+ different options to arrive at the final one or two. Having a rough cut project plan with a timeline, RAID (risks, assumptions, issues, dependencies), and a high level budget is the finishing touch and the mark of a good L3P effort.

Chapter 10 Action Summary

1. Review the layout option and identify (on Post-Its preferably) all key tasks and milestones to get from current state today to the future layout option

2. Once all tasks identified, find key dependencies which can be other tasks, decisions, or events and then work to sequence the tasks accordingly. Post the tasks on flip chart or butcher paper and use arrows to make hard connections to show sequence.

3. Review the tasks and estimate timing of each. Can use a two point time estimate if unsure (low end/high end). Also place an owner for each task

4. Review the tasks and estimate a rough ballpark cost estimate. Bring in some additional people or experts for the discussion if needed. These will be very rough but should be directionally close

5. Summarize the findings – the layout options and how the team go there, the pluses and deltas of the final option (yes, there may still be some deltas), the implementation plan, and the rough cost projection. If multiple options to propose, be sure to estimate timing and cost for both options

6. Report out and decide on direction. Get the chosen layout into CAD with details as soon as possible

Chapter 11: Smaller Scale - Design for Flow (DFF)

John and I were sitting in the airport lounge chatting and sipping some beers. We had just made it to the airport and through security after leaving the plant when my phone rang. It was Ben. The report out had been a few hours earlier and went well. Now it was time to execute and the team had a plethora of activity to do, some of which included reducing the footprints of assembly and machining areas. "Greg and I were talking and think we will need some help with these areas in the plant we 'assumed' would be shrunk. Can you guys help with this? How do we go about it? Does this L3P tool work for smaller areas?" Ben quizzed us as I had put the phone on speaker so John could hear.

"As a matter of fact, we do think we can help. We mentioned it to one of the teams during the week when they were assuming creating cells and shrinking footprints. It's a similar approach as L3P, but scaled down to the cell level. We have been experimenting some more with it and call it Design for Flow or DFF for short," I answered. "We can setup a call next week to go over it with you guys and we can look at the initial focus area," I added.

A few days later, we settled in for the overview meeting, having sent some material to Greg and the team ahead of time. John started the call, "DFF focuses on smaller scale application but is similar in some ways to L3P. It's great if you are looking at a production cell or a department maybe for example. A key benefit now is the ability to actually do some mock ups, which we do with card board or any other materials available. We will essentially design and then try to mock up the cell or space to 'trial' it." I continued, "At the same time, we really take a ground level review of the products and rebuild the process essentially from scratch. We do look at how it is done today, but only to understand. We want the team focused on the product, key characteristics, and building the process to meet the product needs while simplifying for production. We also heavily focus on safety and quality in the new process design, and try to build in defect prevention."

"We really need this in our assembly area. We do a lot of bench assembly and have very low productivity and a lot of quality issues," Jeff, the ME manager commented. "We could save a lot of space likely, probably more than we had assumed, and also improve the process. It could be breakthrough," he added. "We should start with the 4532 line as a pilot. It takes up a lot of space in assembly, is one of the higher volume lines, and causes quality alerts all the time," Greg pointed out.

"Great. Some things we will need include drawings, specifications, annual volumes broken down by part SKU, and if available, the DFMEA and/or PFMEA (Design and Process Failure Modes and Effects Analysis)," I said. "We won't have any DFMEA or PFMEA. The drawings are old and are known to be inaccurate, but we can pull this info together," Jeff replied. "If we don't have a PFMEA, we will create a quick and dirty one during the event. It forms the basis for process design considerations and is important," John commented.

The team pulled together the information as requested, and we went over it in a review call the next week. We had scheduled the event to happen in two weeks, so preparation was critical as there was not much time. John and I used the demand numbers to review takt and prepare for the event. We tasked the team in gathering some card board and other "building" materials to have on hand and asked them to make some decent open space available for mock ups and trials. We had them think through who would be on the team, providing recommendations to have a good cross functional mix including design engineering, product management, production workers, manufacturing engineering, and quality. Having materials people in the event is also advised. We informed them that either design engineering or product management (someone familiar with product) will need to provide the team a functional overview so we all understand the product function and critical aspects.

We wrapped our last review call the week prior. "We will see you next week. It looks as of now everything is in order, so nice work on the prep," I said. "Safe travels. We will see you guys next week," Greg added.

Perspective and Application

So how does the approach change when looking at a manufacturing cell or a department for example? It really doesn't change all that much in principle, just the scale will, and perhaps the detail will increase in some areas since now you are looking at a micro view of a smaller area versus the more macro view of an entire facility. We have crafted a model we call Design for Flow (DFF), which is a good approach to use at a smaller scale for L3P. DFF takes a somewhat similar approach as L3P but is typically aimed at a work cell or more of a value stream area level. It also brings in elements such as standard work, takt analysis, product functional overview, and PFMEA "light" creation if required. This information feeds into the 3P process for the cell. The DFF approach does bring in some detail specific to the manufacturing process design aspects and is a powerful tool for localized and surgical improvements for creating flow and often dramatically improving processes.

The general approach will be similar following a current state analysis, developing key criteria for the design, and then driving team thinking from an innovation perspective to design a future state version. And then using the decision analysis to narrow down and find the best option. It is basically applying the same L3P approach on a smaller, micro level, with less time and rounds than a full L3P run.

If you are dealing with a manufacturing cell or some other similar space, a key difference which can be viewed positively is the ability to do mock ups. Mock ups, also called moon shining in some circles, is using readily available materials such as cardboard to build scaled mock ups of a cell or room so the team can "see" and experience the designed layout physically. These can be two dimensional with cardboard sheets representing the various benches and equipment, or even more powerful, be three dimensional with height and depth brought in as well.

Doing the mock ups is extremely powerful and allows the team to not only see the layout come to life but also allow them to walk through it. If the team is creative enough and puts enough thought and detail into it, they can even do some mock builds and test the flow and even validate standard work and get rough cycle times, which is what we advocate. They can figure out the best spots for materials, tools, and equipment. Think of the power behind being able to do this just after the design was developed on paper. So many times the team has done this and found mistakes or ways to improve the layout considerably.

Cardboard mock ups can be a lot of fun too. Typically, the team gets real creative and lets some good energy out after working with the paper layouts and analysis, and is ready to get after it actually building things. We are always surprised and encouraged by the effort and ingenuity the teams show when doing mock ups. And they really do help. Enough can't be stressed about how helpful it is to actually experience a full 3D mockup of a layout the team worked up on paper. You can then walk through the process to see how good the placement of equipment, materials or tools is in the workspace. You just can't get that from evaluating the paper layout alone, and it is a great approach for cell, value stream, or department layout design.

Chapter 11 Action Summary

1. Identify the space or cell for the DFF application
2. Gather information on the products – drawings, specs, volume info, DFMEA/PFMEA
3. Identify an open space to conduct the event and do mock ups
4. Build a cross functional team comprised of engineering, quality, manufacturing, materials, and production workers
5. Prepare for the event…. Refer to the separate guide on "**Design for Flow**" for a more in depth discussion, application, and execution of the DFF process

Chapter 12: Afterword

No book can make anyone an expert on anything. Books do present knowledge and are great for triggering critical thinking and ideas. Experience is always the best teacher of anything, we firmly believe that. The intent was to introduce some different thinking and application method of 3P concepts applied specifically to the common challenge of designing creative and unique layouts of factories or other spaces to improve space utilization and flow, sometimes under the pressures of time and cost constraints. It's an approach we've used for many years and even today continue to refine the thinking and application of it for maximum effectiveness. We would expect anyone reading this book and applying it to do the same – utilize the basic concepts and adapt to your specific needs. We take the same approach with any improvement tool.

Going through the book, some of the fictitious characters like Greg and Ben were used to demonstrate some of the challenges typically seen when introducing new thinking or tools to people. And this method is different, and those not going through it previously, may tend to challenge it. That's okay. Take them through the process and have them experience it. In doing several of these over the years, we have seen no one go through following the process and say, "that was a waste of time." While you may not generate a total fan club of the approach out of everyone, the majority should see and feel the wisdom behind it.

Much of what was discussed in this book has been covered in the past through experience, both successes and failures. We aren't creators of some new hyped up tool or approach, a lot of these techniques we have molded into this system have been around for some time in everyday life (i.e. Escape Room time challenges) or in continuous improvement circles (i.e. 7 ways). We do feel we have molded the concepts, tools, and principles together to create an effective method that brings both the creative aspects of individual minds and scientific approach of analysis together. The result is a disciplined and rigorous approach that considers a plethora of ideas via divergent thinking and slowly, but methodically refines the concepts through convergent thinking until the best possible design solution emerges in the end.

We hope the book becomes a frame of reference for your layout work, and sparks some thoughts on how to apply in your particular environment and you are able to use it effectively. If you are interested in learning more, specialized tools and templates, getting more in depth support, or just have some general feedback, please reach out to us at info@twoeighty3.com.

Please also be sure to check out and follow my personal blog at www.briansummerfield.com, where you can access other manufacturing and leadership related content as well as supporting tools and materials related to L3P and other topics. Thank you!

Chapter 13: Appendix – References

General Information or Terms

Paper Dolls – Layout technique used with scaled paper cutouts of equipment and assets used to quickly move and position on a to scale blank layout. Paper dolls provide a quick method of developing different alternatives.

8 Wastes – The non-value added work in manufacturing or business processes. They include the following:

1. Defects – Scrap or rework
2. Over Production – Producing more than customers buy
3. Waiting – Time spent waiting for information, parts, etc
4. Non-Utilized Talent or Ideas – Self Explanatory
5. Transportation – Excess movement of materials
6. Inventory – Raw, WIP, or Finished Goods
7. Motion – Excess movement of people
8. Over Processing – Doing more work than customers require

7 Flows – Conceptual view of various flows in manufacturing to consider. Allows us to appraise the current state and help prime thinking for future.

1. Flow of people (motion)
2. Flow raw material (transportation)
3. Flow of WIP (transportation)
4. Flow of finished goods (transportation)
5. Flow of information
6. Flow of equipment (material handling equipment as example)
7. Flow of engineering or tools

7 Ways – Innovation driver, challenges the team to come up with 7 (or more) different alternative options during the design rounds. Sometimes the team achieves it or more, but the point is to strive to get the maximum amount of concepts. It is typically used in earlier rounds during divergent thinking.

Chapter 2

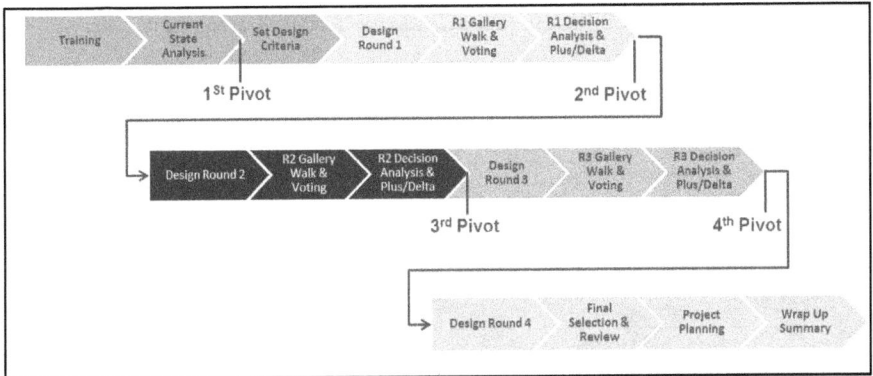

Figure 1: L3P Process Flow Chart

Chapter 5

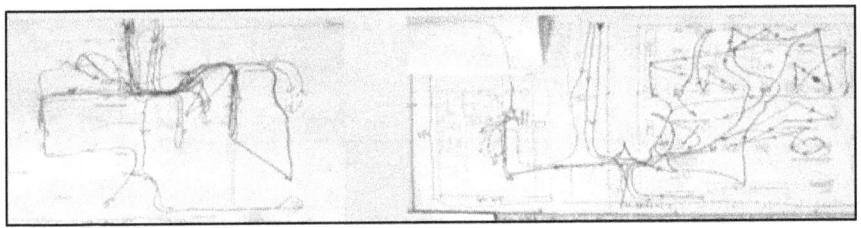

Figure 2: Current State Spaghetti Maps

Chapter 7

3 Votes *Per Criteria* : One (9 - Best), One (3 - Good), One (1 - Okay). Write numbers in boxes - grey shows examples.									
				Layout Option					
Criteria	1	2	3	4	5	6	7		
1									
2									
3									
4									
5									

Figure 3: Voting Ticket

Enter info in GREY shaded cells only															
Enter # of Voters =	12		L3P - Decision Matrix												
Date =	00/00/00		Options												
Dynet - Breckinridge			1		2		3		4		5		6		7	
Hard Limits (Go/No Go)			Go	No Go	Go	No Go	Go	No Go	Go	No Go	Go	No Go	Go	No Go	Go	No Go
Design Objectives	Weight	Score		Score		Score		Score		Score		Score		Score	
VA Space Free	0.30	9.00		1.00		0.50		2.00		1.67		0.50		2.67	
Material Flow	0.40	1.33		2.17		1.17		0.17		0.17		0.50		9.00	
Cost to implement	0.20	0.00		0.00		0.67		0.33		3.00		9.00		0.00	
Time to implement	0.10	0.00		1.17		0.00		0.83		0.00		2.00		9.00	
		9.00		9.00		9.00		9.00		9.00		9.00		9.00	
Total	9.00		3.23		1.28		0.75		0.82		1.17		2.35		5.30
% of Total Score			36%		14%		8%		9%		13%		26%		59%
Ranking		2		4		7		6		5		3		1	

Figure 4: L3P Decision Matrix

Chapter 8

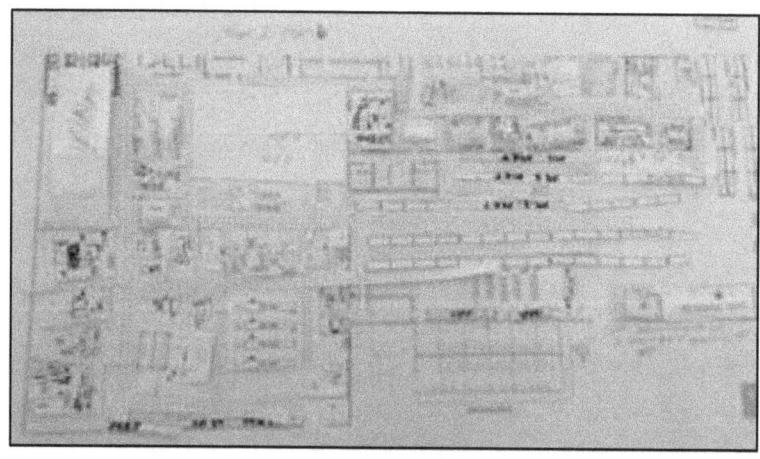

Figure 5: Later Round Layout Concept - Paper Dolls

Page left intentionally blank

www.ingramcontent.com/pod-product-compliance
Lightning Source LLC
Chambersburg PA
CBHW051221170526
45166CB00005B/1989